Microcomputers

Student Monographs in Physics

Series Editor: Professor Douglas F Brewer
Professor of Experimental Physics, University of Sussex

Other books in the series:

Maxwell's Equations and their Applications
E G Thomas and A J Meadows

Oscillations and Waves
R Buckley

Fourier Transforms in Physics
D C Champeney

Microcomputers

D G C Jones
*School of Mathematical and Physical Sciences,
University of Sussex*

Adam Hilger Ltd, Bristol and Boston

© D G C Jones 1985

All rights reserved. No part of this publication may be reproduced, stored in a retrieval system of transmitted in any form or by any means, electronic, mechanical, photocopying, recording or otherwise, without the prior permission of the publisher.

British Library Cataloguing in Publication Data
Jones, D. G. C.
    Microcomputers.——(Student monographs in physics series)
    1. Microcomputers
    I. Title   II. Series
    621.3819'58   TX788.3

    ISBN 0-85274-770-5

Published by Adam Hilger Ltd
Techno House, Redcliffe Way, Bristol BS1 6NX, England
PO Box 230, Accord, MA 02018, USA

Printed in Great Britain by Page Bros (Norwich) Ltd

Contents

Introduction		vii
1	**How the Computer Works**	**1**
	1.1 Data Words	1
	1.2 Codes	1
	1.3 Computer Architecture	3
	1.4 Memory Addressing	4
	1.5 The Processor	5
	1.6 The Instruction Set	6
	1.7 The Input/Output Interface	6
	1.8 Putting it all Together—a Practical Example	6
	1.9 Memories—a Final Word	8
2	**The Input/Output Interface**	**9**
	2.1 Connection to Peripherals	9
	2.2 I/O Addressing	10
	2.3 Communications Interfaces	10
	2.4 Serial I/O	11
	2.5 Synchronous and Asynchronous Signals	12
	2.6 Control Signals and Handshaking	12
	2.7 Interrupts	13
	2.8 Some Practical Peripherals—Keyboard and TV Monitor	14
	2.9 Visual Display Units	15
	2.10 Printers	16
	2.11 Other Peripherals	16
3	**Mass Storage of Data**	**17**
	3.1 The Need for Mass Storage	17
	3.2 Cassette Storage	17
	3.3 Magnetic Disk Storage	18
	3.4 Disk Format	18
	3.5 Disk Reading and Writing	20
	3.6 How much Storage?	21
	3.7 Error Checking	21

4	**Software**	**23**
	4.1 What is Software?	23
	4.2 The Most Primitive Level—Bootstrapping and System Monitor	23
	4.3 Operating Systems	25
	4.4 How do Operating Systems Work?	25
	4.5 Operating System Errors	27
	4.6 Editors and Assemblers	27
	4.7 High Level Languages	29
	4.8 Editors	31
5	**More on Software—Some Notes on BASIC**	**33**
	5.1 Dialects of BASIC; Interpreted and Compiled	33
	5.2 Variable Names and Types	33
	5.3 Interaction with the Operating System	34
	5.4 File Handling	34
	5.5 Interaction with Machine Code	35
	5.6 Graphics	36
	5.7 Using Other People's Software	36
	5.8 Writing Software that Other People Can Use	37
6	**Interfacing to the Real World**	**39**
	6.1 Interfacing to Experiments	39
	6.2 Digital Input and Output	39
	6.3 Analogue Output	41
	6.4 Analogue Input—Hardware	42
	6.5 Analogue Input—Operation	43
	6.6 Other Devices	44

Appendix 1: The 8-bit Word or Byte   **45**

Appendix 2: The ASCII Standard Computer Code   **47**

Appendix 3: The RS232 Standard for Serial Data Transmission   **49**

Appendix 4: Memory Map for the BBC Computer   **50**

Appendix 5: Examples of Simple Operations at Machine Code Level   **51**

Index   **55**

Introduction

A student starting a physics course will almost immediately be confronted by microcomputers doing a variety of jobs, from simple data analysis to controlling experiments. He or she may well have already used a microcomputer in school. Most likely it will have been a different type from the ones in the university laboratory. Certainly it will have had different things connected to it. This book is designed to help that student sort out what the various bits can do, and what they cannot do. In the next few years the microcomputer will become an indispensible tool for any physicist. Just as he or she needs competence with the mathematical and theoretical tools of the trade, so an ability to handle this most versatile physical tool will be necessary.

There are two things that the book will not do. It will not teach the elements of digital logic and the details of computer architecture. Neither will it aim to teach a particular programming language. In the jargon, the book is neither a primer on hardware nor on software. Some excellent books on these subjects are suggested below.

The aim is to give an introduction from the user's point of view, answering the sort of questions that tend to get buried in the more detailed texts. For this reason, the discussion will not be tied down to a particular computer system or language. However examples must be given, and these will generally be referred to the BBC Model B computer and the version of the programming language BASIC used on that system. This is not due to any personal preference, but because this system looks likely to be the one most commonly available in university undergraduate laboratories in the next few years.

Some discussion of operation at a machine language level is essential in order to understand the underlying processes going on. This has not been cast in a 'proper' language, but uses a single phrase or sentence to describe each operation. The aim is to make plain the sequence of operation, without involving the complexities of a particular microprocessor instruction set.

In microcomputer literature, numbers are commonly expressed to the base 16 (hexadecimal base, or hex) for convenience. This can be very offputting at first. When necessary here, numbers are first given in their decimal form, with the hexadecimal version in brackets and marked with a capital H. The reason for using hex will soon become apparent. Numbers in binary form will be self-evident. Appendix 1 tries to make clear this slightly confusing situation.

After reading this book the student should be able to tackle more detailed literature, or the documentation supplied with the system he or she needs to use. But the book and its appendices should still form a useful reference for further work.

References for Further Reading

Computer architecture:

Woollard B 1978 *Digital Integrated Circuits and Computers* (London: McGraw Hill)

Halsall F and Lister P 1981 *Microprocessor Fundamentals* (London: Pitman)

Interfacing:

Sargent M and Shoemaker R L 1981 *Interfacing Microcomputers to the Real World* (Reading, Mass.: Addison Wesley)

Software:

Kemeny J G and Kurtz T E 1980 *BASIC Programming* 3rd edn (New York: Wiley)

Coats R B 1982 *Software Engineering for Small Computers* (London: Arnold)

Apple is the registered trademark of Apple Inc.
CBASIC is the registered trademark of Compiler Systems Inc.
CP/M is the registered trademark of Digital Research Inc.
UCSD Pascal is the registered trademark of the Regents of the University of California.
Z80 is the registered trademark of Zilog Inc.

How the Computer Works 1

1.1 Data Words

Microcomputers store information in the form of binary digits or bits. These bits can take on one of two values which we will call logical one (1) or logical zero (0). In the computer these are represented by voltage levels (usually $+5$ V and 0) but this need not concern us here.

The basic unit of data is a *word* made up of a number of bits. In most microcomputers used at present the data word is eight bits long (although future machines will almost certainly use longer words). This book will deal only with 8-bit data words—usually called *bytes*. Appendix 1 shows how we represent them and gives some examples.

Data words have to be transported from place to place. There are two ways of doing this—parallel and serial. In parallel transmission there are eight wires (plus ground). Each wire carries one of the bits. In serial transmission there are only two wires (signal and ground) and the words are sent one bit at a time. Parallel transmission is obviously quicker (about eight times as fast), but uses eight times as much wire and circuitry at each end. Because of this, inside the computer, where time is at a premium, data words are transported in parallel. Outside, data are usually transported serially.

In both cases, it is obvious that questions of timing and synchronisation are of the utmost importance so that the transmission and reception of data words is not garbled. Inside the computer the synchronisation is controlled by a 'clock' which gives out a continuous signal (usually one or more square waves). This is demonstrated in a simplified form in figure 1.1.

1.2 Codes

Obviously there are 256 different 8-bit words. For someone to transmit information to someone else using 8-bit words the receiver must know what the transmitter meant by a particular word. The information must be *encoded* by the transmitter into a succession of bytes and *decoded* by the receiver.

1

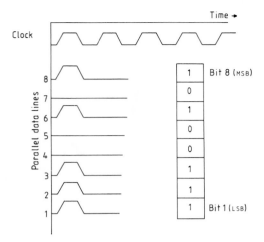

Figure 1.1 Transmission of a data word in parallel form. Note that each line may be permanently at logic 1 or 0. The clock defines the time at which the word is defined.

Within the computer an 8-bit word can be understood in several possible ways. The following two are the most common.

(a) It can be interpreted as a binary number (0 to 255). This is the most obvious code.

(b) It can be interpreted as an instruction to do something. The instruction is despatched to a particular part of the computer and makes this part carry out some operation, depending on the circuitry of that particular part.

Words, and groups of words, can also be used to represent numbers in a more complicated form, or logical functions, but we do not need to deal with such interpretations here.

Outside the computer, in interaction with another computer, or with an operator, the word can be most easily interpreted as a binary number. However it is obviously more useful to have a code which will enable the alphabet and other symbols to be transmitted and received. The common code used by most systems is the ASCII code (American Standard Code for Information Interchange) which is described in Appendix 2.

This code only uses 7 of the 8 bits available, since 128 different words will cover all the Roman alphabet and digits (commonly known as alphanumeric symbols) together with all the common punctuation marks, leaving a few spare. The spare codes are used for so-called *control characters* which will be discussed in Chapter 2. Similarly, the eighth bit is not usually wasted, but is used for checking purposes (see §3.7).

1.3 Computer Architecture

The computer can be broken down into three blocks: processor, memory and input/output (I/O) interface, as shown in figure 1.2. In early computers, each block might have filled a filing-cabinet sized box. Tomorrow, all three might exist on the same tiny piece of silicon. In present-day microcomputers, the processor unit consists of a single integrated circuit (the microprocessor), the memory and I/O interface are both made up of one or more integrated circuits.

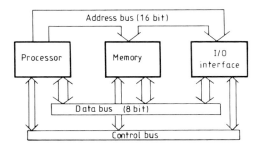

Figure 1.2 Block diagram of computer architecture. Arrows show which way information flows.

The three blocks are connected by three buses (a word taken over from the electrical power industry where *bus-bars* carry power around a factory or laboratory).

The most obvious is the *data bus*, an 8-bit parallel line which enables data words to be transmitted to and from each block.

Second comes the *address bus*. This is a 16-bit parallel line which allows the processor to refer to particular memory elements or I/O ports (see §1.4).

Finally comes the *control bus*. This consists of a number of independent lines. Most of them enable the processor to send signals to control the actions of the other two blocks. Some allow those blocks to start or stop action by the processor.

1.4 Memory Addressing

The memory has the simplest job, so will be described first. It is merely a store for data words, and can be thought of as a large number of mail-slots or pigeon-holes. Each slot (or memory location) has its own address and may or may not contain a data word.

The address of each memory location is specified by a 16-bit binary number which is transmitted along the address bus from the processor. This means that it is possible to have 2^{16} or 65 536 separately addressed memory locations (this number is normally referred to as 64K where K is taken to mean 2^{10} or 1024). In practice, not all the memory addresses may be used, so that the memory size of a particular machine is referred to as 16K (16 384), 32K (32 768) etc. Furthermore, some of the memory addresses may be used to refer to locations which are not part of memory as such, as we shall see later. Some computers, although using a 16-bit address line, employ extra control lines to provide more memory addresses.

There are two ways in which the processor interacts with memory. First, the processor may read from memory. This requires a particular sequence of events, and since this sequence forms the computer's most primitive operation it is worth going over in detail.

An address is placed on the address bus by the processor. Control signals are then sent out by the processor along the control bus which causes the contents of the memory location specified by the particular address to be placed on the data bus. This data word is then taken into the processor and decoded, as we shall see later. Note that the process of reading from memory does not destroy the contents of the memory location, but merely sends out a copy of the contents (see figure 1.3).

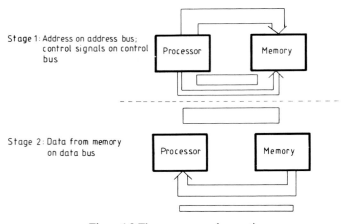

Figure 1.3 The memory read operation.

The inverse process is writing to memory. In this case, the processor puts an address on the address bus and a data word on the data bus, then outputs control codes so that the data word is stored at the address specified. This destroys any word which previously existed at that location.

How the Computer Works 5

1.5 The Processor

Also INTEL, 8080 Series (AMSTRAD)

The processor (or microprocessor) is the heart of the computer. Although there are many types, the vast majority of microcomputers available today use one of two, either the 6502 (Apple, BBC etc) or the Z80 (Sinclair, Tandy). The 6502 is manufactured by MOS Technology and the Z80 by Zilog.

The processor has two jobs: it controls the transmission of data words to and from memory and I/O and it also processes data words inside itself. Each processor has its own built-in set of operations, unique to that type of microprocessor. This set of operations is known as the *instruction set* of the microprocessor.

In the last section the process of reading from memory was described. This is the most primitive example of an operation that a processor can carry out. A particular, and most important, example of that operation occurs when the data word is decoded by the processor as an instruction to carry out some further operation. The processor can only operate on receipt of an instruction from memory, so that the actions of the computer follow the list of instructions read one at a time from memory—this constitutes the computer program.

The operation of the processor can therefore be broken down into a number of steps, each one of which divides into two parts—*instruction fetch* and *instruction execute*. An example of such a step is shown in figure 1.4. This shows the variation with time of the signals on the address, data and control buses. Each

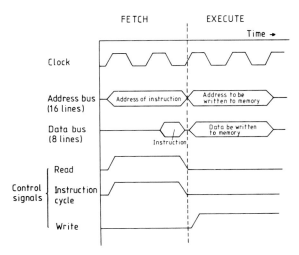

Figure 1.4 The fetch–execute cycle. In the first part of the cycle, an instruction is fetched from memory by a process identical to a memory read, except that the instruction cycle line is at logic 1. In the second part the instruction is decoded and executed. In this case, a memory write is executed.

one of these changes is governed by the clock signal shown on the top line. Both the instruction fetch and the execute part of the cycle take several clock pulses. The time taken by the fetch always remains the same and is governed by the response times of the processor and memory. The time taken by the execute part depends on the complexity of the process to be carried out.

How does the processor know where to get its instructions from? When the processor is switched on, or reset, the address for the first instruction fetch is fixed, either by the processor itself or by the *operating system* (see Chapter 4). For the next fetch the address is normally incremented by one, and so on. In other words, the processor reads a section of memory, one address at a time and interprets the contents as instructions from its instruction set. If the particular section of memory contains random data, the processor nevertheless tries to execute it, usually with disastrous results.

1.6 The Instruction Set

What can the processor do? Briefly, it can take in data, either directly, or indirectly from a specified address. The data can be placed in one or other of a group of registers inside the processor. The registers store data words, just as memory locations do; some processors have three, others have many more. Various mathematical operations can take place inside the processor in a section called the *arithmetic and logic unit* (ALU), using the contents of the processor's internal registers. The contents of the registers can be written to specified addresses in the memory. Also the address of the next instruction to be fetched can be changed. The extent of the instruction set depends on the sophistication of the microprocessor, but essentially it forms a set of primitive operations from which most complex arithmetical and control operations can be constructed. The main features of a typical microprocessor are shown in figure 1.5.

1.7 The Input/Output Interface

This block connects the computer to the outside world. It will be discussed in more detail in Chapter 2, but essentially it enables the connection of devices which can feed data from the outside world into the computer and can send data from the computer to the outside world. Most obviously, it enables input and output to be expressed in a form easily understood by operators, using the ASCII code.

1.8 Putting it all Together—a Practical Example

In this paragraph the general ideas outlined above are applied to a particular

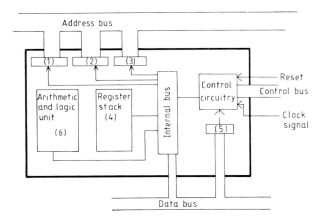

Figure 1.5 The microprocessor. A simplified picture showing the major features. The internal bus enables 8-bit words to be moved from one register to another, to and from the ALU, and also to make up the 16-bit addresses. It also carries control signals. Of the three address registers, the stack pointer is mentioned briefly in §2.7. It carries the address of a section of memory reserved for use by the processor.

The 16-bit registers are: (1) program counter; (2) stack pointer; (3) address buffer. The 8-bit registers are: (4) register stack, which may contain 3(6502) or more registers; (5) instruction register. The arithmetic and logic unit carries out arithmetical and logical operations on the contents of the register stack. The control circuitry decodes and acts upon instructions passed from the instruction register.

computer system—in this case the BBC model B. If the student is using a different system, it would be worth his or her while to study the machine and its handbooks to find out the similarities and differences.

(a) *The processor.* The BBC uses a 6502 microprocessor. This has a clock rate of 2 MHz and therefore a clock cycle time of 0.2 μs. Note that the fetching and executing of an instruction will take a number of clock cycles, depending on the complexity of the instruction. This number is given, when needed, in the data on the instruction set. The 6502 has three internal registers for handling data, plus four others for 'housekeeping' jobs, such as holding the address of the next instruction to be fetched.

(b) *The memory.* Since the 6502 has a 16-bit address bus, in theory the memory could contain 64K locations. However, the amount of memory which can be used by the operator is only 32K, and some of that is used by the operating system (see Chapter 4). Thus only memory locations with addresses 3584 to 32 768 are available. These addresses can be expressed more tidily in hexadecimal base (see Appendix 1) as E00H to 8000H. The other addresses are not wasted, but are used by the computer, as we shall see later.

2.2 I/O Addressing

Since there may well be a number of peripherals interfaced to the computer, the processor must have a way of discriminating between them. This is done by allocating each device an address, just like the allocation of addresses to memory locations discussed in §1.4. The process of outputting a word to a peripheral is then similar to that of writing to a memory location, with one or two small differences.

The obvious way to provide such an address is merely to allocate a block out of the 64K memory addresses to I/O. This is done, for example, in the BBC computer, where memory locations 64 512 to 65 280 (FC00H to FF00H) are used for this purpose (see Appendix 4). This method of addressing (known as memory-mapped addressing for obvious reasons), means that the process of sending a word to an output peripheral is identical to that of writing to memory. The control line which activates the write operation in memory can be used to activate the peripheral and get it ready to receive the data.

However, memory-mapped addressing means a decrease in the number of addresses available for RAM or ROM and the possibility of confusion on control lines. Some microprocessors use a different system, known as I/O mapped addressing. In this case, not all the address lines are used, and the processor has a particular set of instructions which place an 8-bit address on the lower 8 bits of the address line and activate a separate control line, connected only to the I/O devices. This allows 256 separate I/O addresses, which is usually plenty. I/O mapped addressing is more efficient and cheaper in decoding circuitry.

2.3 Communications Interfaces

A peripheral could be wired permanently into the bus system of the computer, but it is more common and more flexible to use an I/O interface chip. Connections from this chip to the outside world then constitute an *I/O port*. Flexibility is gained since these chips are 'semi-intelligent'; they can respond to a range of instructions sent from the processor. For some chips, the instructions merely 'initialise' them, setting them ready for a particular mode of operation. For others, the instructions actually initiate a series of operations by the interface chip itself.

Such chips go under a variety of names, depending on their jobs and level of intelligence. A simple example (in this case, a parallel input/output or PIO chip) is shown in figure 2.2. On the computer side are two registers, each with its own address. The control register accepts a word sent by the processor which defines whether the chip is acting to output data from the computer data bus, or to input data to the bus. This word is part of the very small 'instruction set' of the chip. In the 'output' mode, a data word is sent via the data bus to the data register, ready for transmission to the outside world. In the 'input' mode, a word received at the port is stored in the data register until transmitted along the data bus to the processor. On the 'external' side is an 8-bit port, together with one or more control lines.

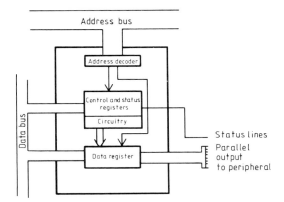

Figure 2.2 Simplified diagram of PIO interface chip. The address decoder activates the appropriate register, depending on the address recieved. The status lines enable the peripheral to control the flow of data from the computer.

In use, the port is initialised by sending the appropriate word to the control register. From then on, the port acts as an input or output interface until re-initialised. In this way one port can act both to input and output data.

2.4 Serial I/O

Other interface chips can be used if the data going to the outside world are in a different form. An example would be the RS232 serial signal described in Appendix 3. In this case, the serial input/output chip (or SIO) would have to do quite a lot. Data can be transmitted in different formats, with or without parity checking (see §3.7) and, most important, at different rates. The transmission rate is defined in bits per second, or in *bauds*. Although the definition of a baud is more complicated, for most microcomputer interfacing one baud can be assumed to equal one bit per second.

For output, the contents of one or more status registers would have to specify the baud rate at which the signal is to be transmitted, and the form of the start, stop and parity bits. Once the data word had been written to the data register, the chip would have to add the extra bits at front and back and transmit the data serially, adjusting the output voltage levels to the RS232 standard (logic 1 between -5 V and -25 V, logic 0 between 5 V and 25 V). Conversely, an incoming signal would have to be stripped of extraneous bits, converted to computer logic levels and assembled in the form of a parallel word before it could be placed on the data bus.

2.5 Synchronous and Asynchronous Signals

The operation of the computer is controlled by its clock. Data could obviously be inputted or outputted at a port in synchronisation with that clock. This is then known as synchronous I/O.

Far more common is the situation where data arrives at a port, or leaves it for a peripheral with no particular reference to the computer clock. This is known as asynchronous I/O. It means, essentially, that each data word is dealt with separately and stored in a register in the I/O interface until dealt with. On output, the clock's responsibility for synchronisation ends when the word is stored in the buffer. On input, the responsibility for synchronisation begins only when the word is read from the registers to the data bus. Because of this, such registers are often called buffers.

2.6 Control Signals and Handshaking

Up to now we have only been concerned with the need to make a single data word acceptable, either to the processor or to the device with which the computer is communicating. However there is a further problem, which can best be seen by giving an example.

A computer is outputting a string of data words to a printer via a parallel interface. The printer interprets the words in terms of the ASCII code and prints the equivalent character.

What decides the speed at which the computer sends words to the output port? In the simplest case, the computer operation might consist of the following steps:

specify address of first data word
read data word from this address
output data word to PIO
increment address
read next data word etc.

Knowing the clock period, and instruction set, it is easy to work out the frequency with which the data words arrive at the port. For example, the BBC computer discussed in Chapter 1 might output data words at a rate of one every 4 microseconds (actually, the rate would probably be rather slower, as extra instructions would usually be added to check whether the end of the data string had been reached etc). Nevertheless, the printer, being a mechanical device would certainly not be able to keep up—common printers print at 100 characters per second or less. So the printer has to be able to send a control signal via the PIO to inform the processor that it is able to accept a new data word, or that it is busy. This is accomplished by a control line from the printer which sets one bit of a register in the PIO. This register has its own address, so that the processor can read the contents of this register and not output a data word until the printer is ready. So the computer operation might look as follows:

specify address of first data word
read data word from this address
read PIO status word (into different register)
if PIO status not OK, go back to last step
otherwise, output data word
increment address etc.

In other words, even the most simple devices connected to the computer may require the processor to test their status before communicating with them. This can be done via the status ports of the I/O interface chip.

The process where control lines are used to start and stop data flow is known as handshaking, presumably from the continental custom of starting and finishing a conversation with a handshake.

2.7 Interrupts

So far we have discussed the control signals required when the processor is 'calling the shots'. For input, on the other hand, the initiative might come from the peripheral. There are two ways by which the peripheral might catch the attention of the processor via the I/O interface.

In the first, the processor is still in complete control. The arrival of data at the I/O chip sets a bit in one of its status registers. Periodically, the processor reads the status register and checks to see if the bit is set. If it is, the processor reads the data from the port. This process is known as *polling*, becuase it means that the processor takes a *poll* of the peripherals.

Polling sounds a somewhat inefficient process. There is another method of getting the same result, via more sophisticated circuitry in the processor itself. This possesses an input control line known as an interrupt line which is normally (let us say) at logic 0. If this goes to logic 1, the processor stops what it is doing. A new, previously specified, address is placed on the address bus, and the next instruction fetch takes place from that address. Instructions at that and the following addresses deal with the interrupt, reading from the peripheral device or whatever is required. This is often known as an *interrupt servicing routine*. The process is equivalent to the action of a telephone bell on someone who is holding a conversation. They break off and start a new conversation with the person on the telephone.

To pursue the analogy, such an interruption can annoy the person conversed with, causing them to lose the thread of an argument. At the very least, the end of the telephone call usually leads to the question 'Now, where were we?'. In the same way, the interrupt must not lead to the loss of data which were being processed or stored in the registers of the microprocessor. So the first instructions of the interrupt servicing routine are often to store the contents of the processor registers temporarily in a block of memory set aside for the purpose, called the *stack*. The address of this is stored in a special register in the processor, called the *stack pointer register*. At the end of the interrupt, the contents of the stack are restored to the processor.

The process of using interrupts is quite complex, but the principle is relatively simple, that a control line from the peripheral, via the I/O port can take control of the action of the processor.

2.8 Some Practical Peripherals—Keyboard and TV Monitor

The most common I/O peripheral system for microcomputers is the combination of keyboard and TV monitor. The keyboard is often mounted in the same box as the computer.

The most primitive type of keyboard, used on very simple computers, has sixteen keys and a 16-to-4 encoder. The encoder produces a unique 4-bit word when one of the keys is pressed, and this 4-bit word can be fed to the data bus when polled by the processor as described above. Obviously each key is labelled with the decimal or hexademical equivalent of the 4-bit word (seen as a binary number).

It would obviously be possible to make a 256-key keyboard to input all the possible 8-bit words, but this would be extremely clumsy. Instead the ASCII code is used, with a keyboard that looks very like a conventional typewriter keyboard, and an encoder unit. To save keys, there is a shift key for upper and lower case letters and for punctuation marks. This idea is extended by having another key, control, which enables the control characters of the ASCII code to be produced (see Appendix 2). The control key merely inverts the seventh bit (b7) of the transmitted word. For example the character G and the control code BEL are represented by 7-bit words identical except for b7. So in a system which implements all the ASCII controls, pressing control–G results in a bell ringing! The data word representing the character is transmitted via a parallel port as described above.

Usually the keyboard is buffered. This means that following the pressing of a key, the appropriate character is stored in a buffer register (part of the keyboard)

The Input/Output Interface 15

until the keyboard port is polled by the processor. The buffer can often hold more than one word and this can cause problems if keys are hit by accident—for example if a key is held down and the keyboard has an automatic repeat facility. When the keyboard port is next polled at a later time, a character may already be present in the buffer which may cause unexpected results.

Output, on the simplest computers, consists of some device which can display a character. A data word is outputted via a port to a buffer (or latch) where it is held until replaced—since the visual signal obviously needs to be held longer than a few clock cycles, and the processor has other things to do than continually feed the signal to the display. The word is then decoded and displayed in one of a number of forms, of which the simplest is a pair of hexadecimal digits.

On most present day systems, the appropriate ASCII symbol is displayed on a TV receiver or a TV monitor. This is by no means a simple job. The TV set makes up its picture by scanning an electron beam across a screen and varying the intensity of that beam. It scans in a preset pattern, to make a flicker-free picture, rather than simply from top to bottom. For the computer to use this, it is necessary to specially reserve and set up an area of memory to copy the screen, making allowance for the scanning pattern. This is known as mapping the screen in memory. ASCII symbols to be displayed have to be encoded to the correct places in the screen map. For an output display, the contents of this section of memory have to be written to a video output port where they are converted into a serial signal at appropriate voltage levels to drive the TV monitor. If a TV receiver is used, a further stage has to take place. The signal has to be converted into modulation on a UHF carrier so that it appears to the TV set to be identical to a radiated TV signal.

If colour TV is used, the principle is the same, except that further information has to be supplied to get the correct colour pattern. Of course, all the procedures required are worked out in advance by the computer manufacturer and form part of the computer's operating system (see Chapter 4).

2.9 Visual Display Units

Very often, the I/O functions of keyboard and monitor are combined in a single unit separate from the computer. This is known as the *visual display unit* (VDU).

The VDU is usually interfaced to the computer via a standard RS232 serial port; baud rates have to be set and handshaking lines have to be connected correctly for both computer and VDU. The keyboard and buffer unit functions exactly as discussed in §2.8. However, the VDU takes care of all the work involved in screen mapping. The computer merely outputs an ASCII symbol and the VDU does the rest, and displays the symbol on its screen.

It is usually possible for the computer to make the VDU carry out more complicated operations. For example, the position at which the next symbol will be displayed is marked by a cursor. By outputting the appropriate combinations of ASCII characters to the VDU, the cursor can be moved round the screen.

VDUs can be used in two modes, full duplex and half duplex. In half duplex, any character hit on the keyboard is displayed on the screen by the VDU itself. In full duplex, the character is merely transmitted to the computer and the responsibility of displaying the character rests with the computer. In normal operation, the full duplex mode is used and the computer 'echoes' back the character. In other words, when a VDU is used in full duplex mode, and hitting a key does not result in anything on the screen, the fault may well lie with the computer or with the interface, not with the VDU.

2.10 Printers

Some of the problems involved in using printers have already been discussed in §2.6. A printer is a vital peripheral for any serious computing. Space does not permit a general discussion, but the most common types rely on printing characters by using a matrix of pins, each of which can produce a dot on paper—hence they are known as dot-matrix printers. Electric typewriters and daisy-wheel printers produce a better typeface but are much slower. The printer interface may be serial or parallel. Obviously if a printer is to be operated at some distance from the computer, serial is preferable, otherwise there is little to choose between them.

A printer usually contains an input buffer of at least 1K words. If its print speed is 100 characters per second, this means that the computer can output a message of fewer than 1000 characters in a second or two, finish its output routine, and go on to other operations. The printer can then carry on and finish printing independently. Since printer speeds are limited by mechanical considerations and the cost of memory is likely to decrease, printer buffers are likely to get larger which will cut the computer time involved in printer output.

Most printers are also semi-intelligent and will alter character size, typeface etc on receipt of the appropriate control characters from the computer port.

2.11 Other Peripherals

There are many other more specialised peripherals, which cannot be discussed here. One which may be encountered is a Modem, an interface which enables data to be transmitted via the telephone system. However the class of peripherals involved in bulk storage of data are important enough to have the next chapter devoted to them.

Mass Storage of Data 3

3.1 The Need for Mass Storage

It would obviously be impractical to have to input via the keyboard every program or file of data each time the computer is switched on, so there must be some form of back-up store for such data. We have seen in §1.9 that such storage can be provided by ROM. However, once programmed, an ROM cartridge cannot be easily reprogrammed and does not provide an easy way of storing data and programs after an hour's or a day's work with the computer. At present, such a facility is provided by magnetic media in one form or another. A thin layer of a magnetic oxide forms a very good medium for storage of digital data. Digital signals can be easily recorded and reproduced. A large number of words can be stored in a relatively small area, and the information is not easily destroyed, provided reasonable precautions are taken. There are two forms of magnetic back-up storage in general use at present: cassette and disk.

3.2 Cassette Storage

The common audio cassette system provides the cheapest means of storing data. Standard cassette recorders and cassettes may be used, although somewhat better results can be obtained using specially designed systems.

How are the data recorded on the cassette? To start with, it can be seen that recording is in serial form, so that a special serial interface is required. However, the data can be recorded in many ways, the simplest being merely by setting logic 1 at some high voltage and logic 0 at some low voltage. Depending on the baud rate of the serial interface, the data are stored as a series of pulses. More common is the so-called 'Kansas City' standard where logic 1 is represented by 8 periods of a 2400 Hz square wave and logic 0 by 4 periods of a 1200 Hz square wave.

The Kansas City standard gives a data transfer rate of 300 baud, which is typical. However, a quite modest program or data file might contain 4000 ASCII characters. This means that to transfer such a file to or from cassette would take over two minutes. Such a time may be acceptable if the transfer is only done once, on switching on; but many applications require regular copying to and from storage during a program. In such a case, not only is the time delay prohibitive, but the serial nature of tape storage makes the job almost impossible. To reach data at the end of a tape, the whole tape has to be played through (although short cuts can sometimes be taken). Baud rates could be increased, of course, but the cassette tape winds at a constant rate, so a higher baud rate means less tape area per bit of data and a consequent increase in possible errors due to flaws on the tape. Recorder head misalignment is also common. This causes errors and, more annoying, makes it often impossible to transfer data from one computer to another via different cassette players.

In general, although cassette storage is cheap, it is very limited.

3.3 Magnetic Disk Storage

Magnetic disks use the same physical method of recording data as does the cassette, but get over most of the problems of access to data mentioned above. Two varieties are common—hard disks and floppies. Hard disks come in sealed units, usually integral to a computer system and will not be discussed further; however many of the principles of operation of floppy disks apply to them.

A floppy disk, as the name implies, consists of a plastic disk, coated on one or both sides with magnetic material. The disk is enclosed in a protective sleeve from which it is never removed. The surface of the disk is visible through a slot in the sleeve. The disk is operated in a disk drive. This has two essential mechanical parts, a spindle which rotates the disk at high speed and a read/write head which moves along a radius of the disk and makes contact with it through the slot in the sleeve. The head operates just as the cassette head does, writing binary digits onto the magnetic disk surface or reading them off it. Drives can have two heads, one for each side of the disk, enabling the use of double-sided disks.

The original floppy disks were 8 inches in diameter. Today the most common size is $5\frac{1}{4}$ inches (sometimes called minifloppies or diskettes). Smaller and cheaper disks with greater storage capacity are likely in the future. The great advantage of the disk is that by moving along its radius arm the head can quickly access any part of the disk. How this is done is the next topic to be discussed.

3.4 Disk Format

Data cannot be written at random on the surface of the disk. There has to be a certain well-defined format so that input and output can be quick and efficient. Unfortunately there are many varieties of disk format, but the large majority of them share the same features (see figure 3.1).

Firstly the head can move along the radius in a number of steps. These steps define the number of tracks on the disk. 8-inch disks have 77 tracks per side. Diskettes may have 35 or 40 tracks per side. Each track is split up into sectors and an index hole marks the beginning of the first sector. The division into sectors is carried out by writing a data pattern onto each track of the disk, a process known as formatting. This produces so-called 'soft-sectored' disks. 'Hard-sectored' disks exist with an index hole marking the start of each sector, but they are much less common.

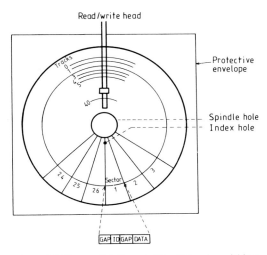

Figure 3.1 Floppy disk format. The disk spins within a protective envelope. A notch in the envelope enables the disk to be made 'read only', so that vital files are not deleted by mistake. On 8-inch disks the slot must be covered for normal 'read-write' use. On 5-inch disks, the slot must be left uncovered.

Each sector is divided into four parts. First comes a gap, comprising a number of bytes of junk (usually FFH), then come a number of bytes which identify the sector (the ID block). These include side number, track number, sector number, and some check digits. Another gap is followed by the block of data bytes. Depending on the format, this block may contain 128, 256 or some other number of bytes and some check digits. The ID block enables a particular track and sector to be identified and read from or written to.

For 8-inch disks there are two standard formats, with 26 sectors per track in each case. In FM or single-density format, each sector holds 128 data bytes. In

MFM or double-density format, each sector holds 256 bytes. In the second case, data are recorded in a form which enables twice as many bytes to be packed into a sector—hence 'double density'. There is no standard format for diskettes and most computer manufacturers have defined their own format. For example, the BBC uses 40 tracks per side, 10 sectors per track, 256 data bytes per sector.

3.5 Disk Reading and Writing

Sending data to and from disks is obviously complicated. Fortunately most of the hard work is done by another semi-intelligent interface chip which contains a number of registers, each with its own I/O port address. As in other I/O interfaces, there is a data register, and a control register which can accept and interpret instructions from the processor. However, there are also track and sector registers, possibly other control registers, and, most important, a status register which stores information about the status of the read/write head at any particular moment.

The chip can carry out a number of operations. The most important are:

seek track—move head to track specified in track register
read sector—read data from sector specified in sector register
write sector—write data to sector specified in sector register
write track—write to every location on a specified track.

A typical routine for reading data from a particular sector into memory might go as follows, with the processor writing the appropriate data words to the addresses of the appropriate registers.

Write track number to track register.
Write sector number to sector register.
Write 'seek track' instruction to control register.
Wait until status register shows that head has arrived at the right track.
Issue 'sector read' instruction.
Wait until status register shows that head has found the right sector.
Read word from I/O chip data register.
Write word to memory address.
Increment memory address.
Check status register to see if whole sector has been read in.
If whole sector not read in, read next word from data register.
Etc.

How long will this process take? This will obviously depend on the type of disk drive, number of tracks etc. However, typically it might take an average of 200 ms for the head to find the track and 'load' (make contact with the disk surface). Then the sector must be found, which might take some 50 ms. The time required to read the sector could be about 10 ms, so that the total time to read a sector of 128 bytes would be 260 ms. Thus the time to read a 4000 character file would be under 10 s. In fact, most of the time is taken in moving the head to the correct track. If our 4000 bytes were stored in adjacent sectors on the same track, the reading time would be nearer to 2 s. Part of the job of disk operating systems, which we will discuss in Chapter 4, is the storage of data in sectors so that read and write times are minimised.

The process of writing to a sector follows the reading process exactly, with obvious differences. The 'write track' instruction enables disks to be formatted. A pattern of all the bytes on a track is set up in memory, including gaps, ID records etc. The 'track write' then enables this to be written on the disk. This process has to be carried out whenever a new disk is used and is carried out by a special 'format' program.

3.6 How much Storage?

Given the information in the last few sections, it is easy to work out how much data can be stored in any particular case. A C15 cassette, if used in Kansas City format, could store about 27 Kbytes—but it would take 15 minutes to read that 27 Kbytes into memory.

A single-density 8-inch disk can store 256 Kbytes per side, and a double-density, double-sided disk, 1 Mbyte. The capacities of diskettes vary according to their format. In general, a figure of around 100 Kbytes per side for single density and 200 Kbytes per side for double density is usual. However, not all the disk is necessarily available for users. The first two tracks are often reserved for operating system use, including a directory of the files stored on the disk (see Chapter 4). Finally, for storage of really large amounts of data, hard disk units can store 5 Mbytes or more.

3.7 Error Checking

Transmission of data to and from peripherals always carries the danger of errors creeping into the signal. Because of this, error checking routines are built in. These cannot be gone into in detail, but there are two common ones.

Firstly, parity checking. In the ASCII code, only 7 bits are used, leaving one spare in each byte. If the signal is degraded on transmission, the most likely result would be a logic 1 being read as logic 0, or vice versa. Parity checking enables this to be spotted. Before a byte is transmitted, the individual bits are added together. This must result in 0 or 1. Bit b7 in the data word is then adjusted so that the total always comes to 0 (even parity) or to 1 (odd parity). At the reciever, the individual bits are again totalled. If the result gives the wrong parity, the data have been corrupted. Most peripherals and interfaces can be adjusted to check for odd or even parity or not to carry out the check.

In transmission to or from disk units, the parity check method is not applicable, as 8-bit words are transmitted, which may or may not represent ASCII characters. In this case *cyclic redundancy checking* (or CRC) is carried out, which involves summing all the data bytes and expressing this sum in a form that can be checked on read or write.

As might be expected, major errors in mass storage media come via electrical or mechanical damage to the disk surface. A major cause of electrical damage is the switching on or off of the system with a disk loaded. Power supply 'glitches' can often be transmitted to the disk read/write head, and hence destroy data on the disk surface. However, with reasonable care, disk storage media are relatively reliable and stable.

Software 4

4.1 What is Software?

The last three chapters have discussed the physical components which go to make up a computer system. These components are normally referred to as *hardware*. We saw how the various components operated, and the important fact that they could be controlled via the processor to carry out complicated sequences of operations. The sequences of instructions devised to make the hardware operate are known as *software* and the technique of production of such software is known as software engineering.

4.2 The Most Primitive Level—Bootstrapping and System Monitor

In §1.5 we saw how the operation of the computer consisted of the execution, one at a time, of data words read from memory which the processor interpreted as parts of its instruction set. In Chapter 2 we saw that the processor could control peripherals by a similar process of writing data words to their semi-intelligent interfaces.

The question remains, how do the data words get stored in memory in the first place? In the most primitive computers, data words could be placed directly into memory. A set of keys connected to the address bus enabled a memory address to be set manually, and a similar set of keys connected to the data bus could set up a data word. Activation of the memory-read control placed the word in memory. This can still be done on simple microcomputers, but it is fairly easy to get the processor to do the job itself.

When the computer is switched on, or the reset line to the processor is activated, a starting address (usually 0000) is set on the address bus and an instruction is fetched from this location. The trick is to have a program already existing in the machine which will enable the operator to input data easily from a keyboard and store the data in memory. Such a process is called bootstrapping (or booting) as the computer starts itself up and gets itself organised—pulling itself up by its own bootstraps.

The program is stored in ROM and is usually known as a *monitor* (not to be confused with the hardware peripheral TV monitor). At its simplest, the monitor does three things. It first outputs to the I/O port (TV or VDU) a prompt character to indicate that it is in operation. It then causes the processor to poll the keyboard until one of a very limited range of characters is detected. Detection of one of these characters then activates a further set of instructions to carry out one of a number of operations (see figure 4.1).

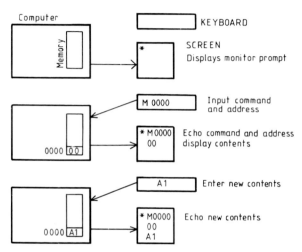

Figure 4.1 Monitor operation. *Step 1*: prompt character output to screen. *Step 2*: Input of command character, followed by memory address. Command and address echoed, followed by output of memory contents. *Step 3*: Input of new contents, contents echoed.

To take an example, assuming a simple hexadecimal keyboard with a few extra keys (see §2.8). The character S might cause the following set of operations.

The next four digits keyed are interpreted as a memory address.
The contents of that address are displayed on the screen.
The next two digits keyed are interpreted as an 8-bit word, and stored at that address.

The character G might cause the following operations.

The next four digits keyed are interpreted as a memory address. The next instruction fetch takes place from that address.

These two simple commands enable 8-bit words to be written to memory and executed by the processor. The monitor program which does the job is quite complex and might involve over a hundred instructions. Most monitors have a rather wider range of commands and are therefore more complex (see for example the Demon monitor described by Sargent and Shoemaker).

The monitor enables the preparation, input and running of programs at the most primitive level, with 8-bit words keyed into the computer in binary or more often in hexadecimal code. This is known as machine code, because the operator is dealing directly with the instructions handled by the processor.

4.3 Operating Systems

The job of a monitor is to handle the 'housekeeping' jobs required to input and execute strings of instructions at the lowest level. However, in more complex systems there are many other housekeeping jobs which can be better done by the computer. Interfacing to other peripherals, such as printers, needs instruction routines to initialise etc. Most important, the interfacing to disks and other storage media needs a repertoire of routines to handle storage, and to keep a record of which data have been stored where. To carry out such housekeeping requires a far more extensive version of a monitor, known as an *operating system*.

Since the job of an operating system is to handle the computer operation down at the 'nuts and bolts' level, it obviously has to be tailored to the particular machine it is working on, and even to the particular peripherals connected to that machine. This is the job of the computer manufacturer. For example the BBC computer has an operating system Acorn DOS (standing for disk operating system). However there are other operating systems which are portable, and can be used on many types of machine. The two best known on microcomputers are the CP/M and UCSD p-system. To the user, any computer using (say) CP/M looks the same; but someone (hopefully the manufacturer, but possibly the user) will have had to tailor the system to the particular collection of hardware employed. The advantage is that any data file or program produced on a CP/M system should be able to be read by any other system—so long as both systems format their disks in the same way (see §3.4).

4.4 How do Operating Systems Work?

To start with, the operating system must be bootstrapped when the computer is switched on. In some machines, such as the BBC, the whole operating system is held in ROM. In others, including most machines which use CP/M, the operating system is held on one or two tracks of a disk (known as a system disk). In that case, there is a simple bootstrap routine in ROM which reads into memory the contents of the system disk tracks. In many cases the booting is a two- or three-stage process, with the ROM routine reading in a bootstrap from disk which reads in and activates the rest of the operating system. In any case, the operating system occupies a considerable part of the computer memory, reducing the space free for the user (see figure 4.2 for examples).

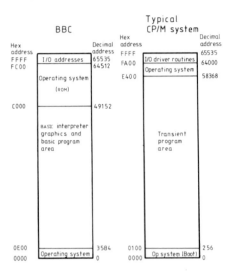

Figure 4.2 Memory Usage by two typical operating systems. In the BBC Acorn DOS the operating system is stored in ROM. In CP/M, the operating system is stored on disk. On reset, the bootstrap loader is read into locations 0000H to 0100H. It then loads the remainder of the operating system.

As with the monitor, the first job of the operating system is to set up the interfaces so that the operator can converse with the computer. However, the range of commands that can be accepted by the computer is much wider. Most are concerned with the storage and recovery of data on the disks.

Data are stored in *files*. A file may contain a program in a high level language, a program in machine code, data, text, or merely junk. To the operating system they are all just blocks of words which have to be stored in sectors on the disk and kept track of. Each file is allocated a name and its name and location is stored in a directory, usually on track 2 of the disk. Usually a file type code is also allocated, together with other information, depending on the complexity of the operating system. For example, the CP/M operating system allows the user to allocate any name of up to seven characters to a file, and also a three-character file type code. Thus, as we shall see in §4.6, a 'family' of files may be produced of different types, each with the same filename.

In the BBC operating system (Acorn DOS), on the other hand, a filename only is usually specified, although particular special files may be designated by single letters.

Each operating system has its own repertoire of commands which enable the contents of specified sections of memory to be stored as a file, or a particular file to be written into memory. Other commands enable files to be renamed and deleted (which merely means removing the file name from the directory). Commands can also display the disk directory, transfer a file to a printer port, or carry out other operations.

4.5 Operating System Errors

Since errors are bound to occur when blocks of data are moved around, the operating system will have routines to deal with the most common faults and to inform the operator when an error has occurred. Some errors are trivial, such as attempting to write to a disk which has been protected from further writing, or trying to write to a disk that is already full. Others are less trivial, such as when a flaw in the disk surface has been detected by CRC checking (see §3.7). Data from damaged disks can sometimes be saved by special recovery programs.

On detection of an error, a message is sent to the VDU or screen. However, since the routines are not able to cope with every eventuality, such a message is often unhelpful and sometimes misleading.

4.6 Editors and Assemblers

Operating systems usually include the simple monitor commands in one way or another. However, the tedious job of piling up hundreds of hexadecimal digits can be eased by most operating systems. The major problem of writing sequences of instructions in machine code is that it is necessary to remember what each byte means, and keep some reference as to what goes where. This is done most easily by first writing a program, not in machine code but in so-called *assembly language*. In this, each machine code instruction is allocated a *mnemonic*. It is then much easier to write programs in terms of mnemonics (for example, for a 6502 microprocessor, the word E8H causes the X register to be incremented; it is far easier to remember the mnemonic INX).

Originally, therefore, programmers wrote programs in assembly language then translated them into machine code. Today, rather than writing on paper, then converting by hand, the assembly language can be written into the computer and then translated by the computer into machine code.

The first part of this process is carried out by an *editor routine*. This merely takes in and stores in a file the ASCII characters which make up the assembly language program. (Editors have, in fact, a much wider range of uses—see §4.8.) Once such a file has been produced, a further routine, known as an *assembler*, converts the mnemonics into the machine-understandable machine code. The editor and assembler are usually referred to as *utilities of the operating system*. The difference is that while the editor can be used to produce any sort of text file, the assembler has the single job of converting mnemonics specific to a particular type of microprocessor into machine code which is also specific to that type of

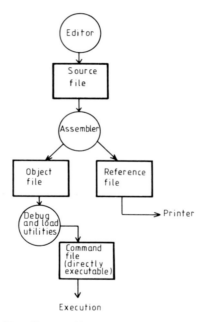

Figure 4.3 The edit–assemble process. The assembler usually produces two files, the object file which contains the machine code, and a reference file which gives both a source code and the appropriate machine code and memory locations. The load utility converts the file into one which will be directly executed on receipt of the filename. In the CP/M system the process produces a family of files:
  Source – FILENAME.ASM
  Object – FILENAME.HEX, Reference – FILENAME.PRN
  Command – FILENAME.COM.

microprocessor. Thus the Z80 has its own instruction set and its own assembly language mnemonics which are completely different from those of the 6502.

Of course the mnemonics are only unique to the microprocessor in the sense that they are generally agreed by the people who produce the assembler utilities. It would be quite possible to have two completely different sets of mnemonics referring to the same machine code instructions. In fact the Z80 processor is a development of an earlier design (the 8080). The Z80 instruction set includes the 8080 instruction set plus a lot more. Because of this, Z80 mnemonics are completely different from those used for the 8080. However, a program written in 8080 mnemonics and then assembled using a 8080 assembler will run perfectly well on a Z80! (Conversely, a program written in Z80 mnemonics and assembled with a Z80 assembler would run on an 8080 so long as only mnemonics corresponding to 8080 instructions were used.)

The file of mnemonics is often referred to as the *source code*, as compared to the machine-code file which is referred to as the *object code*.

In theory, the result of the edit/assemble process is a file which can be read into memory and be executed by the processor. In practice, the assembled file will almost certainly contain errors (or *bugs*) which means that it does not make the computer do what the operator wants. In order to find such faults, *debugger* utilities are provided which enable the machine-code instructions to be executed one at a time, to find out what is going wrong. Such a process is known as single-stepping the machine-code program. The debugger usually contains facilities to change individual instructions to get the program running correctly. The debugged routine can then be stored as a new object file.

Finally, a utility is provided which converts the object file into a file which can be recognised by the operating system as a machine-code program to be executed by the computer. Thus the process of editing and assembly produces a family of files which represent different stages in the process which is summed up in figure 4.3.

4.7 High Level Languages

The processes of editing and assembly make the production of software less burdensome. But each operation has still to be spelt out into a series of primitive steps, making even such a simple job as multiplying two numbers together incredibly wearying. Obviously, the next stage of sophistication is to make each mnemonic of source code correspond to a whole set of primary operations in machine code (the mnemonic is then more usually referred to as an instruction). This produces a so-called *high level language*. The source file is produced as before, but the process of converting into machine code is considerably more complex. This process is known as *compiling*, and the utilities which do the job as *compilers*. The process is shown in figure 4.4.

A major problem with the edit/compile system is due to the fact that nearly every first draft of a program contains errors. These are of two sorts. Syntax

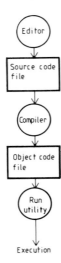

Figure 4.4 The edit–compile process.

errors are errors in the source code which mean that the assembler or compiler cannot interpret the source (a simple example might be the use of a mnemonic which does not exist in the particular assembly language.) These are easily picked out during the compiling process. More common are run-time errors—actual mistakes in the processes which the processor has been told to carry out. In a compiled system, such errors can only be found at the end of the edit/compile process when the software is tried out. Corrections can only be made by re-editing the source file and re-compiling—a process which might take some minutes for each minor correction or change.

A way out of this is the use of an *interpreted* language. In this case, the source file can be produced as before, but the processes of compiling and running the program are carried out together by a utility known as an interpreter. On setting the program running, each instruction of the source program is individually compiled into machine code and acted upon. This makes the operation of an interpreted program very much slower than a compiled one, but means that corrections to individual sections of the source file can be made and the program re-run immediately. Because of this, such systems are sometimes known as *interactive* (see figure 4.5).

The software required to interpret a high level language is going to be complex and fill a large amount of memory. Often it is stored in ROM. For example the BBC BASIC interpreter takes up over 16K (see Appendix 4). If the computer is dedicated to using a particular high level language such as BASIC, the operating system usually sets up the interpretation process on booting up and a prompt which requests instructions in that language is the first thing displayed by the computer. In other cases, the interpreter may have to be read from disk and run like any other machine-code file.

There are numerous high level languages. In physics laboratories, those most likely to be used are BASIC (standing for Beginners All-Purpose Symbolic Instruction Code), PASCAL (named after the mathematician) and FORTH. For better or worse, BASIC is going to be the most common language for some years so we shall spend the next chapter discussing some points of importance for users.

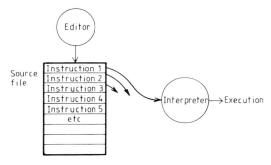

Figure 4.5 The edit–interpret process. The interpreter reads each high level instruction, converts it into a series of machine code operations and executes them.

4.8 Editors

It should be fairly obvious that the editing processes described in §§4.6 and 4.7 are essentially identical. In fact the editor utility is a flexible one with wide application. An editor can enable the production of mnemonic files, source-code files for any compiled language which uses ASCII characters, or even text files such as letters, memos or books. In fact, the software of a word processor is merely an editor utility with a few extra features added.

There are two general types of editor. The first, referred to as a *line-oriented editor*, enables the production of files one line at a time. Each line can be looked at and modified individually. The second type are usually called *screen editors*. These enable a screen-full of the file to be displayed at a time and allow changes to be made at any point on the screen.

Line editors date from the time when the usual device for interacting with the computer was a teleprinter, a sort of modified typewriter which could only display one line at a time. Although line editors are acceptable for the production of assembly language files, screen editors are far superior for any other application.

The CP/M operating system has its own line editor called ED. However there are many screen editors which can be used with the CP/M operating system. The

BBC has its own editor which is something of a cross between a line and screen editor. This is fine for editing lines in BASIC programs, but not much use for text files which can be produced and edited via word processor routines available as plug-in ROM packages.

More on Software— Some Notes on BASIC 5

5.1 Dialects of BASIC; Interpreted and Compiled

There are many versions of BASIC which differ from the original version which is sometimes called 'Dartmouth BASIC'. It is not the aim of this chapter to teach the language, but merely to point out some features of importance in different BASIC dialects, which the reader might come upon.

As mentioned above, BASIC is an interpreted language. It is possible, however to get BASIC compilers which (in theory) give the advantages of both interpretation and compilation. The program is first written and tested ('debugged') in the normal interpreted fashion, then saved on disk as a source file and compiled, either into object code or into some intermediate code. A 'Run' utility then enables the compiled program to run. In theory, this should speed up the operation of the program by a factor of 10 or more. In practice, the increase in speed is not so great. In most programs the majority of execution time is taken up by I/O operations such as disk read and write (see §3.5). Compilation will, of course, make no difference to the time taken by these operations. A compiled BASIC file will, however, take up much more space in memory than the original source file. This may cause problems if memory space is limited.

Note that the process of producing a BASIC program and saving it on disk in the usual fashion is not the same as using an editor to produce a text file which consists of a list of BASIC instructions. In the first case, the BASIC instructions are usually stored in an abbreviated or 'tokenised' form, and such a file can be loaded and run in the normal way. The text file, on the other hand, is merely a text file to the computer. However a BASIC program stored in a text file can be run using a special operating system command (*EXEC in Apple DOS). This treats the text file as though it were being typed in one instruction at a time from the keyboard.

5.2 Variable Names and Types

Different dialects of BASIC let programmers name variables in different ways. BBC BASIC lets variable names consist of any number of letters or numbers, subject to a few restrictions, such as not starting with characters identical with a BASIC instruction (for example, PRINTOUT would be an illegal name). In Applesoft, on the other hand, only the first two characters are read, so that variables 'Voltage' and 'Volume' in a BBC BASIC program would be read as the same variable 'Vo' in Applesoft.

Most dialects use three types of variables; string, integer and real number. String variables are the simplest, merely consisting of a string of ASCII codes, and are stored in memory as such. Integer variables are stored as binary numbers in consecutive memory locations. In the BBC, four words are used per integer, giving a maximum integer size of $\pm 2\,147\,483\,648$ (2^{31} with one bit for the sign). In the Apple, two words are used, giving $\pm 32\,767$. Finally, each real number has to be stored in a number of memory locations in the form of an integer and its corresponding power of 10. In both BBC and Apple, five words are required, organised so that numbers between $\pm 1.7 \times 10^{38}$ can be used.

The precision is defined by the number of digits assigned to the integral part (or mantissa). For BBC and Apple, numbers can be stored to a precision of nine significant figures. If precision greater than this is required, for example when subtracting two nearly equal numbers, special techniques have to be used. Other dialects of BASIC, such as CBASIC work to higher precision.

Obviously, the higher the precision, the more words have to be used to make up each number and so the more complicated and slower the variable handling routines. Conversely, calculations are usually executed more quickly if integer variables are used, rather than reals.

5.3 Interaction with the Operating System

At first sight, when it comes to producing code to calculate a particular function, the various dialects of BASIC are not too different. However major differences occur when input and output operations occur, since at this point the operating system of the particular computer has to be used. Essentially, what is required is that the BASIC interpreter has to transfer control to the section of the operating system which is needed (for example, the disk read operation, if data is to be loaded from disk). In most cases, operating system commands can be embedded in the BASIC code and are handled by the interpreter (for example in BBC BASIC such commands begin with an asterisk *). Often, BASIC instructions can be used which modify the operation of the operating system. This is usually done by changing the contents of particular memory locations in the operating system by instructions such as 'Poke' or '?'. The operation of such vital commands can only be worked out from the manual of the computer and with practice.

5.4 File Handling

File handling techniques are often the most difficult to master, as the operating system has usually not been devised with a first-time user in mind. A file containing a BASIC program is usually easily handled by instructions such as 'Load' or 'Save'. More difficult are data files which contain a large number of individual items of data which may be needed by the program during operation. Such files are usually referred to as *random access files*. To set them up, the size and number of data elements has to be specified, and some form of pointer established to define where each element of data is kept. Once the data have been read into the file, the pointer enables a particular record to be read into memory. When data from a file have to be read in, the file has to be opened by a special instruction. After use, the file has to be closed again, or an error will result when an attempt is made to open another file—even if this occurs in some later, separate, program. This is because opening and closing of files is carried out by the operating system, not specifically by the BASIC program. There is usually an operating system command (such as 'Break' on the BBC computer) which 'tidies up' after any program, including closing any files that have been left hanging open.

In general, the best way to handle such data is to load them into memory in a block, to form a so-called 'look-up table' to which the program can refer. This can be done right at the beginning of the program with a simple file-read routine. To use a look-up table is obviously very much quicker, and avoids the difficulties of opening files and searching for individual records in the middle of operation.

5.5 Interaction with Machine Code

The compiling of a BASIC source file is one way to speed execution of the program. However, it is often the case that only certain sections of the operation need to be speeded up (see §6.5). In this case, the most obvious choice is to produce such a section in machine code, store it in memory and call it up from the BASIC program. The major problem with this technique is that two levels of programming are getting confused. At the BASIC level, the interpreter takes care of all the mundane jobs of allocating memory space etc. In fact the person programming in BASIC need never know where the program is stored in memory, where the variables are kept etc. But a machine code routine needs to have its memory location specified, and this must obviously not overlap with any space used by the BASIC interpreter, or the operating system. This requires some study of memory maps such as Appendix 4, and careful adjustment of interpreter parameters.

A space in memory having been found, the machine code routine can be put together using operating system utilities, as in §4.5, or otherwise. In BBC BASIC for example, it is possible to call up an assembler from within a BASIC program, so that assembly language mnemonics can be included within the BASIC file. On running, assembly takes place when the appropriate point in the program is reached and the machine language routine is stored in a chosen area of memory, ready to be used later in the program.

5.6 Graphics

Each computer has its own graphics routines. Essentially, a portion of memory has to be allocated to emulate the screen, taking account of the problems discussed in §2.8. The highest resolution graphics takes up a lot of memory space, so that various graphics options are usually provided (table 5.1). Handling graphics routines to provide graphs etc is usually relatively easy, although it is also easy to spend a considerable amount of time producing labelled axes, etc. Because the process of assembling a graphic picture in memory is relatively slow, there are various ways of speeding it up for use in animated displays. These go by various names such as 'Player-missile graphics' or 'Sprites', and are beyond our scope.

Once a graph has been produced, it is usually necessary to reproduce it in hard copy form on a printer. This is not an easy job, because the interlacing of screen lines makes direct copying of the graphics section of memory impossible. Routines which do the job are called *screen dump utilities*. Note that such a utility may be designed specifically for a particular type of printer, using control codes specific to that printer. Since the number of points per row of a dot matrix printer is fixed (at 560 for an 80 character 7×9 dot printer), the graphics mode to be dumped may also have to be specified, so that the program can somehow or other squeeze the appropriate number of horizontal points onto one line of print. Similarly, since the distance moved by the printer paper on 'line feed' is usually fixed, the vertical axis of the graphics display may be squeezed or stretched. This will not matter for ordinary graphs, but may well mean that a circle on the screen will come out a very flat ellipse on the printer.

Table 5.1 Graphics modes on the BBC computer (Model B). The screen can be thought of as divided into a number of cells, each of which can be defined as black or white (or a colour if a colour monitor is used). The greater the number of cells, the higher the resolution.

Mode	Resolution†	Colours	Memory required
0	640×256	2	20K
1	320×256	4	20K
2	160×256	16	20K
4	320×256	2	10K
5	160×256	4	10K

† Resolution = number of cells on X axis × number of cells on Y axis.

5.7 Using Other People's Software

Just as no one would expect to have to produce their own operating system for their computer, it would be stupid to have to write programs for the sort of common jobs for which computers are used in the laboratory, such as curve fitting. Producing such programs (known fashionably as 're-inventing the wheel') may be a useful exercise, or may be needed for special cases, but most students in physics laboratories will be using software which has either been bought, or which has been produced in the lab for a particular purpose. Some software will be self-contained, engaging the student in a dialogue by asking for data and producing the results. Other utilities, such as a straight-line fitting and output routine, might be in the form of a routine to be inserted in a particular program. It might even form a BASIC file in its own right to be linked to other programs (for example, by the BASIC command 'Chain').

In whichever way pre-written software is used, there is one vital procedure to follow. Always 'back up' the program. If you are given some software on a disk, immediately copy it to your own disk and hand the original back, or store it away safely. This is a vital rule, as it is fairly easy to destroy a disk-full of software. If you have a back-up copy, this does not matter. The rule applies just as much when software is being developed. Save a copy—even if only a partial copy—on disk (or on tape if necessary) at very regular intervals. Many people have spent hours working on a program, only to lose the lot by a mistake, or even worse by someone inadvertently switching off the computer.

On copyright of software; if the software is given to the student by a lecturer or supervisor, it should be assumed that copyright questions are dealt with by the institution. On software acquired in other ways, the question is still obscure legally, but the same rules probably apply as apply to the copying of commercial music records and cassettes.

5.8 Writing Software that Other People Can Use

A major weakness of BASIC as a high level language is that it enables very badly structured programming. By this is meant the fact that programs can be written as long strings of instructions with very little pattern to show from a listing as to what is actually going on. To try to sort out what is happening in someone else's BASIC program can be very similar to trying to unravel a badly knotted ball of string. Languages such as PASCAL force programmers to structure things in a tidier and hopefully more comprehensible way.

However, BASIC programs can be structured tidily, especially in dialects such as that used on the BBC, and it is vital that such methods are seen as central, not just as optional extras.

Writing software that other people can use is a huge subject, the surface of which cannot even be scratched here, but there are a couple of points that are of particular importance to physics students.

Many programs will require the input from the keyboard of large amounts of data. During that input, it is quite likely that someone will make a keying error. It should not be necessary to key the whole lot in again. Similarly, it should be possible to check the data before proceding to the next stage of the program. At later stages of the program, it may be necessary to provide extra data, or define a parameter. If an error is made here, it should not 'crash' the program, meaning that it has to be started again from scratch. The program should announce what it is doing in detail at any particular stage of operation. Where there are options, a *menu* should be presented, including an option to go back to an earlier stage or stages.

Such suggestions may seem obvious, but when a student is wrestling with the production of routines to solve particular mathematical problems, solving equations, drawing graphs and the like, they tend to get overlooked. These few paragraphs can in no way replace a good software engineering course, but may serve to highlight points of importance in the physics laboratory.

Interfacing to the Real World

6.1 Interfacing to Experiments

Chapters 2 and 3 dealt with the interfacing of a microcomputer to the most common peripherals. In a laboratory, the computer will possibly be also interfaced to a range of rather simpler devices connected to experiments, to control them and/or to analyse data. In such cases the computer may have to deal with digital signals, analogue signals or both.

Digital output might mean the closure of a switch, or the lighting of a lamp. Digital input might involve monitoring when a switch is closed or checking when a light beam has been broken.

Analogue output might involve producing a signal to be displayed on an oscilloscope, or generating a waveform to drive a device or a loudspeaker. Analogue input might be the result of monitoring any time-varying voltage signal, for example from a photodetector, a microwave receiver or a discharging capacitor.

6.2 Digital Input and Output

In principle, digital input and output are very easy to deal with, given a parallel I/O port of the type discussed in §2.3. Such a port gives us eight independent digital lines for input or output, and some PIO chips even allow each line to be addressed separately for either input or output. To output a digital signal, all that has to be done is to connect up one of the lines (for example b1, the least significant bit (LSB)) to our external device. Then an instruction to output any word which has b1 at logic 1 (for example 11111111 or FFH) does the job.

However, in real life things may be a little more difficult. To start with, the signal arriving at the external device will only last the equivalent of one clock pulse (0.5 μs for a BBC). The output process can be repeated, of course, but the computer may well have better things to do. So some external circuitry will probably be needed to latch the signal (go to logic 1 when instructed, and stay there). Also, only a certain maximum current can be drawn from the PIO chip without nasty things happening to it, so that if a relay is to be closed or a light emitting diode (LED) lit, some amplification stage will have to be added.

Similarly, digital input is relatively simple, with the processor reading the appropriate input line to the PIO (note that most PIOs interpret an open circuit at the input port as logic 1). If input is taken from other circuitry, there is also a danger of damage to the computer if excessive voltages, or radio frequency noise, appear at the input port. Because of this, inputs are usually isolated by a device known as an optoisolator which combines an LED and a phototransistor. The incoming signal lights the LED which then activates the phototransistor. The process sounds clumsy, but the computer is completely electrically isolated from the input. Figures 6.1 and 6.2 show examples of such simple input and output.

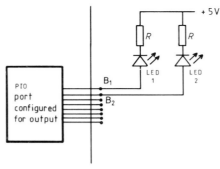

Figure 6.1 Digital output. In this case, lines b1 and b2 are connected to two LEDs. Output of any word with b1 and/or b2 equal to 0 will light one or both of the LEDs (typically, R might be 200 Ω).

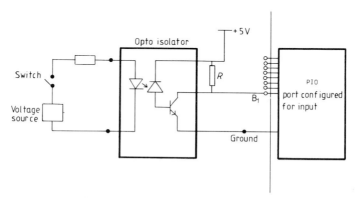

Figure 6.2 Digital input with optoisolator. The voltage source may have any value, consistent with the parameters of the O/I unit. Switching on the input LED, causes the second diode to conduct, hence b1 of the input is pulled low.

6.3 Analogue Output

The next job is to enable the computer to produce analogue signals—signals that can vary continuously. This can be accomplished via our parallel port using a fairly simple circuit (figure 6.3). This circuit provides an output of zero volts when the word 00000000 (00H) is sent to the port and 10 V when the word 11111111 (FFH) is sent. In general, the D–A (digital-to-analogue) converter (or DAC) produces its analogue output in a time shorter than the clock time. However, it does not produce a continuously variable signal, but one which changes in step fashion. Since the input word is one of 8 bits, the step size will be 1/256 of the maximum voltage.

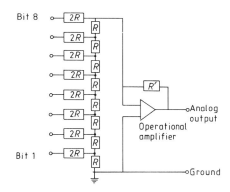

Figure 6.3 Digital–analogue convertor. The operational amplifier merely acts as a buffer. R may be of the order of 10 kΩ. If R' is set equal to R, logic 1 (5 V) on all pins produces an output of 5 V.

These facts tend to restrict the use of the DAC as a waveform generator. To generate waveforms, software can be used to control the output to the PIO. For example, if a string of data words is sent to the output port, starting from 0 and incrementing by one each time, a sawtooth waveform will result (note that the contents of a register will repeat cyclically so that FFH $+1=$ 00H). The period of the sawtooth will equal 256 times the time taken to carry out the sequence of machine code instructions:

output register contents to port
increment register contents.

For a 6502 processor as used in the BBC computer, this means that the maximum frequency of the sawtooth will be around 1.3 kHz. Of course, if lower resolution were acceptable (e.g. counting in steps of 2), higher frequencies could be obtained. More complex waveforms could be obtained by a variety of software techniques, such as using a look-up table (see §5.4).

In many computers, special chips are used to produce sounds, controlled from the processor using high-level language instructions.

6.4 Analogue Input—Hardware

The process of converting an analogue signal to a form which can be handled by the computer is more complicated than the inverse process of digital–analogue conversion. It may be carried out in a number of ways but only the two most important will be discussed here.

The first and cheapest method uses a DAC and comparator (see figure 6.4). The computer outputs a 'trial' signal via the DAC and this is compared with the input signal. If the input is greater than the trial signal, a larger signal is tried; if smaller, a smaller signal, until a match is found. In other words, the computer carries out a search process until it generates a signal equal to the analogue input. The most efficient search process is one where the area to be covered is split into two on each trial (this is known as a binary search). Given that the computer has an 8-bit data word, nine trials will be needed, on average, to do the conversion. Each trial will involve a sequence of processor operations of the form:

output test word
input comparator signal
compare with zero
if not zero, set up new test word.

Thus, using a 6502 processor as before, the conversion process will take some 70 μs. This puts a limit on the rate of change of signals to be examined.

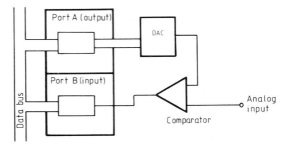

Figure 6.4 Search–compare analogue–digital conversion. A two-port PIO, DAC and comparator are used. The search compare procedure is carried out by the processor.

The process can be speeded up by using a dedicated analogue-to-digital (ADC) chip. This contains all the circuitry needed to carry out the search, and sometimes also its own clock generator which controls the search routine. The chip acts like a processor which can run only one program. It is connected to a parallel port like any other peripheral, a clock signal is provided either from the computer or from a separate oscillator, and the conversion process is initiated by a control signal from the processor. Such chips can cut the conversion time to about 10 μs.

Another type of ADC uses a different process. This is known as flash conversion, because it takes place in a flash (relative to search–compare). If search–compare might be thought of as a serial method of conversion, flash conversion is the parallel equivalent. The analogue signal is simultaneously compared to a number of equally spaced reference voltages. The comparator outputs are then encoded in such a way as to give the equivalent binary number. This process can take as little as 50 ns, giving the possibility of analysing signals varying at frequencies up to 20 MHz. The snag is that for n-bit conversion, $2^{(n-1)}$ comparators are needed, which makes flash ADC units much more expensive than search–compare units.

6.5 Analogue Input—Operation

Two considerations limit the use of ADCs to monitor the outputs of laboratory experiments; precision and speed. An 8-bit ADC can obviously only measure to a precision of 1 in 256 (to put it another way, the maximum resolution is 1 in 256). Note that to obtain such a resolution the signal has to be amplified or attenuated so that its largest value is just equal to the maximum input acceptable to the ADC.

Such a resolution (better than 0.5%) is usually acceptable for a wide range of experiments. However, if a signal varies over several orders of magnitude, and changes at the lower end of the range are of interest, the input amplitude will have to be altered to give good resolution over the amplitude range of interest.

The response time of search–compare ADCs was quoted above as 10 μs. If this is slower than the time required to carry out the necessary machine language routine to initiate conversion and read the result, the frequency response is limited by the ADC itself. This would not be true if the ADC input were read via an interpreted language such as BASIC, since in this case the interpretation process would usually take much longer than the conversion process. (In the BBC, in fact, the time required by the shortest machine code routine is about 10 μs.)

If flash ADCs are used, the response time is limited by the clock rate of the processor. Thus for the BBC computer, using a 6502 processor, working at a clock rate of 2 MHz, the maximum sampling rate for data will be about 100 000 samples per second. It can be shown that this means that only signals which change at a frequency less than 50 kHz could be accurately monitored by the computer.

To put the last statement in more precise terms; only those signals in whose spectrum the great majority of energy is concentrated at frequencies well below 50 kHz will be faithfully reproduced in the computer memory.

Obviously, if resolution is sacrificed, by using fewer than 8 bits, speed can be increased or cost decreased. Similarly, if speed is sacrificed, larger words can be used and resolution increased. There are also techniques for directly sending data from ADC to memory without having to go via the processor (known as direct memory access or DMA), which get round the limitation of clock speed. These are beyond the scope of this book.

Analogue-to-digital conversion will be used to monitor a signal which varies with time, and to store successive values of that signal in memory. If the input from the ADC is an 8-bit word, this can easily be done in machine code, with one memory location for each signal value. If the software is in BASIC, the values can be stored as integer variables with minimum wastage of memory space. Once stored, the data can be displayed in the manner of an oscilloscope trace as a function of time, or can be analysed.

Many microcomputers have their own built-in ADC circuits. The BBC, for example, has four analogue input channels. Each of these has 12-bit resolution (1 in 4000 or 0.025%), but the response time is 10 milliseconds per channel.

6.6 Other Devices

One other device which must be mentioned is the stepper motor. This enables mechanical devices to be controlled by the processor. Essentially, a type of electric motor is used which will make a fractional revolution in response to a digital signal. For example, a stepper motor is used to drive the read/write head along a radius bar in a floppy disk drive. The incremental movements of the motor define the tracks on the disk (see §3.3). A detailed introduction is beyond the scope of this book, and students are referred to Sargent and Shoemaker.

Appendix 1:
The 8-bit Word or Byte

b8	b7	b6	b5	b4	b3	b2	b1

Each bit can take the value 0 or 1.

Ways of representing on paper:

(*a*) In terms of a binary number, e.g. 10101000 (= 168 in decimal arithmetic).

(*b*) In terms of two hexadecimal digits (numbers to the base 16). Divide the byte into two 4-bit 'nibbles' and express each separately, e.g. 10101000 = 1010 1000 = A8H.

The convenience of this method can be seen from the conversion table below showing conversion from binary to decimal to hexadecimal.

```
Base 2:   0 1 0 1 0 1 0 1 0 1 0 1 0 1 0 1 0 .
          0 0 1 1 0 0 1 1 0 0 1 1 0 0 1 1 0 .
          0 0 0 0 1 1 1 1 0 0 0 0 1 1 1 1 0 .
          0 0 0 0 0 0 0 0 1 1 1 1 1 1 1 1 0 .
          : : : : : : : : : : : : : : : : :
Base 10:  0 1 2 3 4 5 6 7 8 9 10 11 12 13 14 15 16 17 18 19 ...
Base 16:  0 1 2 3 4 5 6 7 8 9 A  B  C  D  E  F  10 11 12 13 ...
```

To convert a multidigit hexadecimal number to decimal, use the table overleaf.

Column							
4		3		2		1	
hex	dec	hex	dec	hex	dec	hex	dec
0	0	0	0	0	0	0	0
1	4096	1	256	1	16	1	1
2	8192	2	512	2	32	2	2
3	12288	3	768	3	48	3	3
4	16384	4	1024	4	64	4	4
5	20480	5	1280	5	80	5	5
6	24576	6	1536	6	96	6	6
7	28672	7	1792	7	112	7	7
8	32768	8	2048	8	128	8	8
9	36864	9	2304	9	144	9	9
A	40960	A	2560	A	160	A	10
B	45056	B	2816	B	176	B	11
C	49152	C	3072	C	192	C	12
D	53248	D	3328	D	208	D	13
E	57344	E	3584	E	224	E	14
F	61440	F	3840	F	240	F	15

Example

Consider the 8-bit word: 01010000. We can write this for our convenience as 50H. It may be understood inside the computer as:

     (*a*) an instruction (depending on the particular microprocessor);
or (*b*) a binary number (=80 in decimal);
or (*c*) a character in ASCII code (in this case 'P').

Appendix 2: The ASCII Standard Computer Code

The 8-bit word is defined as shown below:

MSB							LSB
b8	b7	b6	b5	b4	b3	b2	b1

The *least significant bit* (LSB) is b1 and the *most significant bit* (MSB) is b8. In the ASCII code, b8 is not used and may be used for parity checking (see §3.7). If a particular device implements the full ASCII code, the control codes activate a particular operation. For example, BEL causes a bell to ring, LF causes paper in a printer to be moved up one line or the cursor on a VDU to be moved down one line.

			b8	0	p	p	p	p	p	p	p
	bits		b7	0	0	0	0	1	1	1	1
			b6	0	0	1	1	0	0	1	1
			b5	0	1	0	1	0	1	0	1
b4	b3	b2	b1								
0	0	0	0	NUL	DLE	SP	0	@	P	`	p
0	0	0	1	SOH	DC1	!	1	A	Q	a	q
0	0	1	0	STX	DC2	''	2	B	R	b	r
0	0	1	1	ETX	DC3	#	3	C	S	c	s
0	1	0	0	EOT	DC4	$	4	D	T	d	t
0	1	0	1	ENQ	NAK	%	5	E	U	e	u
0	1	1	0	ACK	SYN	&	6	F	V	f	v
0	1	1	1	BEL	ETB	'	7	G	W	g	w
1	0	0	0	BS	CAN	(	8	H	X	h	x
1	0	0	1	HT	EM	)	9	I	Y	i	y
1	0	1	0	LF	SUB	*	:	J	Z	j	z
1	0	1	1	VT	ESC	+	;	K	[	k	{
1	1	0	0	FF	FS	,	<	L	\	l	|
1	1	0	1	CR	GS	−	=	M	]	m	}
1	1	1	0	SO	RS	.	>	N	∧	n	~
1	1	1	1	SI	US	/	?	O	—	o	DEL

Control characters

NUL	Null	DLE	Data link escape
SOH	Start of heading	DC1	Device control 1
STX	Start of text	DC2	Device control 2
ETX	End of text	DC3	Device control 3
EOT	End of transmission	DC4	Device control 4 (Stop)
ENQ	Enquiry	NAK	Negative acknowledge
ACK	Acknowledge	SYN	Synchronous idle
BEL	Bell (audible or attention signal)	ETB	End of transmission block
BS	Backspace	CAN	Cancel
HT	Horizontal tabulation (punched card skip)	EM	End of medium
		SUB	Substitute
LF	Line feed	ESC	Escape
VT	Vertical tabulation	FS	File separator
FF	Form feed	GS	Group separator
CR	Carriage return	RS	Record separator
SO	Shift out	US	Unit separator
SI	Shift in	DEL	Delete

Appendix 3: The RS232 Standard for Serial Data Transmission

(a) Form of serial signal (1200 band).

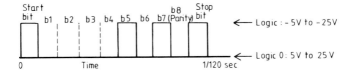

(b) Connection for serial output (e.g. printer).

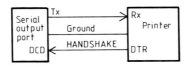

(c) Connection for input/output (e.g. terminal).

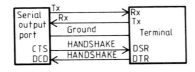

Note: the handshaking lines are not standard, and may vary from one piece of equipment to another.

Appendix 4: Memory Map for the BBC Computer

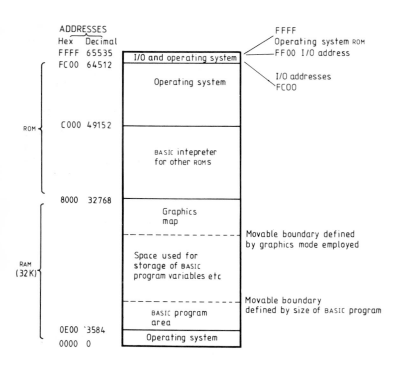

50

Appendix 5: Examples of Simple Operations at Machine Code Level

To demonstrate the ideas discussed in Chapter 1, this Appendix gives an example of a very simple operation inside the microcomputer, and how it is executed by the two most common microprocessors.

The operation involves taking a block of 256 data words which has been stored in memory at some previous time, and sending them, one at a time, to an output port.

In words, it can be broken down into the following primary operations.

(1) Initialisation: storing starting address of block; storing index for counting.

(2) Operation loop: read from memory location to accumulator; write from accumulator to port; increment address; decrement counter; check if counter is zero; if counter not zero, return to start of operation loop; if counter zero, continue to next stage.

(3) Finish: end program.

Because of their different architecture and instruction sets, different processors will implement this operation in different ways. In each case, the instructions required (from a much larger set) are tabulated and discussed. Then the necessary program is assembled.

6502 Processor
Part of instruction set. The operation is described, a mnemonic is given for it, the actual contents of memory are given (in hex for clarity) and the number of clock cycles for the operation to be carried out (fetched + executed). Symbols $\langle p \rangle, \langle n \rangle$ and $\langle m \rangle$ are used to indicate values of bytes which will have to be chosen by the programmer. The 6502 has three internal registers, A, X and Y.

File data, 26
  directory, 26
  handling, 34
  random access, 34
  type, 26
Format, double density, 20
  single density, 20

Glitch, power supply, 22
Graphics, 36

Handshaking, 12
Hardware, vii
Hex, vii
High level languages, 29
Housekeeping, 7
  op. system, 25

I/O BBC, 7
  parallel, 10
  serial, 11
ID block, 19
Initialisation, 11
Input, isolation, 40
Instruction, execute, 5
  fetch, 5
  set, 5
Interface, I/O, 6
  printer, 16
Interfaces,
  communications, 10
Interfacing, 39
Internal registers, 6
Interpreter, BASIC, 30
Interpreters, 30
Interrupt, line, 13
  servicing, 14
Interrupts, 13

Kansas City standard, 17
Keyboard, ASCII, 14
  BBC, 7
  buffer, 14

Language, BASIC, 31
  FORTH, 31
  interactive, 30
  PASCAL, 31
Latch, 14, 39

Machine code, 51
Machine language, vii
Magnetic disk, 17
  media, 17
  storage, 17
  tape, 18
Mass storage, 17
Memory, addressing, 3
  BBC, 7
  DRAM, 8
  location, 3
  non-volatile, 8
  PROM and EPROM, 8
  RAM, 8
  read, 4
  response time, 8
  ROM, 8
  static RAM, 8
  volatile, 8
  write, 4
Memory map, 24
Memory map, BBC, 50
Mnemonics, 27
  6502, 29
  8080, 29
  Z80, 29
Modem, 16
Monitor system, 23
  TV, 8, 14

Object code, 29
Operating system, 6
  systems, 25
Output, analogue, 41
  LED, 40
  relay, 40

PIO, 10
Parallel transmission, 1
Parity checking, 21
Peripherals, 9
Polling ports, 13
Port, I/O, 10
Printer, interface, 16
Printers, 16
  daisy-wheel, 16
  dot-matrix, 16
Processor, 5
  6502, 5

Processor, *continued*
  BBC, 7
  Z80, 5
Program, 5

RS232, 11
RS232 standard, 12, 49
Receiver, TV, 15

Screen mapping, 15
Screen dump utilities, 36
Sector, hard, 19
  ID, 19
  soft, 19
Serial transmission, 1
Signals, asynchronous, 12
  synchronous, 12
Software, vii, 23
  back up, 37
  copyright, 37
  engineering, 23
  menus, 38
  other people's, 36
  structuring, 37
Source code, 29
Square wave, 1
Stack, 14
Stack pointer, 14
Stepper motor, 44
Synchronisation, 1
Synchronous signals, 12
System, disk, 26

Table, look-up, 35

UCSD p-system, 25
UHF output, 15
Utilities, op. system, 28

VDU, 15
Video output, 15

Word, 1
  8-bit, 1
  interpretation, 2